*Delicate colors of western
hemlock sapling, Lane County.*

*Front Cover: Second growth
Douglas-fir forest photographed near
Table Rock, Clackamas County, 1993.*

*Rhododendrons and bear grass flourish in a meadow surrounded by
young Douglas-fir near Lolo Pass, Mt. Hood National Forest.*

Oregon's New
FORESTS

A Photographic Essay by
STEVE TERRILL

Interpretive Text by
PATRICIA K. LICHEN

OREGON FOREST RESOURCES INSTITUTE

SPECIAL THANKS TO:

Bill Swindells of Willamette Industries for his "inspiration;" Jim Brown, Fred Robinson
and the Oregon Department of Forestry, Dean George Brown and the Oregon State University College
of Forestry, and Dr. John Beuter of Duck Creek Associates, Corvallis, for their comments
and suggestions; Cathy Baldwin, Curt Copenhagen, Cary Jones, Greg Miller, Dave Odgers, Dick Rohl,
Shep Tucker, and Chris West for their review; the Board of Directors of the Oregon Forest Resouces Institute:
Jim Bradbury, George Brown, John Byrne, Ken Everett, Rob Freres, Jr., Craig Hanneman,
Russ McKinley, Jennifer Phillippi, Jeff Van Natta, and Brad Witt; OFRI Staff: Leslie Lehmann,
Mike Beard, and Roberta Taylor; Oregon's forest landowners who made their property readily accessible to
the photographer; and, those Oregonians who have led the way in protecting forest resources through research,
education, conservation activism, and enlightened forest management.

Unless otherwise specified, the photographic images in this publication depict private forest land
that has been previously harvested. We are grateful to the following companies and agencies
for allowing access to their land, and for their assistance in this project: Boise Cascade Corp.,
Crown Pacific, Freres Lumber Co., Inc., Georgia-Pacific Corp., Longview Fibre Co., MAP, Inc.,
Ochoco Lumber Co., Oregon Department of Forestry, Oregon State University
College of Forestry, Rough & Ready Lumber Co., Starker Forests, Inc., Stimson Lumber Co.,
Sun Studs, Inc., Van Natta Brothers, Warm Springs Tribes, Weyerhaeuser Co.,
Willamette Industries, Inc., and Willamina Lumber Co.

Produced by Companion Press
Santa Barbara, California
Jane Freeburg, Publisher/Editor

Designed by Lucy Brown
Illustration on p. 3 by Linda Trujillo
Color Separations by Photolith Systems
Printed by Riddle Press, Beaverton, Oregon

ISBN 0-944197-39-6

Contents

Top: Unfurling ferns.

*Center: Ponderosa bark, stripped
and eaten by forest animals.*

*Below: A horse fly lights on
an evergreen branch.*

Opposite: Wild blackberry vines.

Sword ferns on forest floor of second growth Douglas-fir stand,
near Crabtree Creek, Linn County.

Opposite: Lush vegetation along
Camp Creek, Grant County.

A WALK IN THE WOODS

A visit to Oregon's forests invigorates the senses. Depending on which side of the mountains you walk, you might smell damp earth or dry grass; see overwhelming variations on the color green or the reddish trunks of ponderosa pines; feel dripping rain or hear the squeak of snow beneath your boots. Dulled senses awaken in a forest, coerced by the sights, scents, and sounds of the living land.

A walk through a forest renews, soothes, and inspires. If you're walking in a second growth forest (one which has been cut and has regrown) you might be surprised by its beauty and diversity. The management of Oregon's forests has changed in response to advances in science and evolution in understanding. What hasn't changed is our need for wood.

*I*n our hectic, mechanized lives, it is easy to forget how much we depend on forests. They sustain us physically: a roof over our heads and floor beneath our feet; warmth from the woodstove; a desk and the stationery to write a letter; huckleberries for a pie; a chair to rock a baby to sleep; music from guitars, pianos, and violins; even the very oxygen we breathe. Look around you, wherever you are now. What do you see that came from the forest?

Forests also give us the opportunity to ski, hike, camp, hunt, and fish; and intangibles such as a sanctuary for spiritual renewal and a place to make a connection with nature. Every day, in a variety of ways, we continue to depend on forests.

Almost half of Oregon is forest, about 28 million acres. Just over 19 million of these acres are capable of growing commercial timber. On the wet, western side of the Cascade Range, over 11 million acres of timberland grow Douglas-fir, hemlock, cedar, spruce, alder, and maples. On the dryer, eastern side, Oregon boasts another 8 million timberland acres of ponderosa, lodgepole and sugar pine, along with larch, Douglas-fir, Engelmann spruce and true firs. A moist climate and fertile, volcanic soil make Northwest Oregon one of the best places in the world to grow trees: a seedling planted just fifty years ago in the Oregon Coast Range may stand 120 feet tall and 18 inches in diameter today.

Fog softens a meadow scene of ten-year-old mixed conifer saplings;
mature second growth Douglas-fir loom in the background.

HUMAN HISTORY IN THE OREGON WOODS

Oregon's people have an undeniable bond with their forests. Through the years, attitudes towards forests have changed, but those who live here have always had a connection with the woods. From the Native Americans who wore the inner bark of cedar trees, through the early loggers who believed the forests were limitless, to the urbanites who head for the woods on the weekend, we have used the technology of our time to meet our needs. People of the Wasco and Warm Springs tribes used fire to keep trees from overrunning their berry fields, and to create openings for game. Early loggers laboriously sawed through mighty trees with 12 foot long "misery whips," and shipped the wood south to support California's gold rush. Mills sprang up in response to the housing boom after World War II, and today's high-tech, computerized mills continue to meet America's growing demand for wood.

Forests—the scrawlings of insects, the heat of fire, the ramblings of animals, and processes of death and life—would continue without our intervention. However, because we need and desire all that trees become, we have learned how to encourage their rapid growth. Like other animals, we take what we need from the woods—but our ability to influence the forest far exceeds theirs. We walk a narrow trail between promoting the rapid growth of trees and allowing nature's processes to continue at nature's pace. Oregon's past century of logging has shown we must—and can—continue to revise management practices as we better understand the forest's complex cycles.

Evergreen logs stacked on a "log deck" at Warm Springs.

UNDERSTANDING THE FOREST ECOSYSTEM

Oregon's landscape has changed considerably from the early 1800s when Meriwether Lewis and William Clark explored here. Their journal reports, "The trees of larger growth are very abundant; the whole neighborhood of the coast is supplied with great quantities of excellent timber." But even in those early days, the Northwest was not covered by a continuous mantle of old growth. As trees within a forest died from fire or were blown over in windstorms, new plant life thrived in the opened, sunny area. As this process continued through decades and centuries, old growth and open areas "moved" throughout a forest. Some of the trees that were saplings during Lewis and Clark's visit are today's old growth. But much of the forest the explorers saw

Alder, fallen logs, and ferns line a forest stream.

Tillamook State Forest
Burned in a series of fires, 1930s–1940s
Although the forest was replanted with
conifers, these alder grew naturally.

Little North Fork River
Tillamook County
Photographed February, 1992

has made way for social and economic development, cut to meet the needs of generations from settlers to suburbanites.

When it became obvious that the woods were not limitless, natural regeneration gave way to tree planting and active forest management. More recently we've learned less obvious lessons. A forest ecosystem is so complex we may never completely understand it, but years of field research have revealed many of the secrets of a healthy forest. These lessons are now being applied in Oregon's managed forests.

For years, fallen trees were routinely removed from forest streams because biologists assumed they impeded fish migration. But research showed that salmon, steelhead, and trout use pools created by the logs as resting and spawning areas. They also need trees and shrubs along the stream banks to cool the water, reduce sediment in streams, and supply a home for the insects that fish eat. As a result of this increased understanding, logs are now left in the streams, and trees along the banks are managed to enhance fish habitat.

And dead trees are no longer routinely cleared away to make room for young seedlings. Scientists discovered that dead trees are needed in a healthy forest. Standing dead trees, called snags, harbor insects that various animals eat, and their soft wood makes it easy for birds and other animals to create cavities to live in. Fallen trees on the forest floor provide homes for insects and small animals, and retain water that is slowly released during dry periods. As the tree decomposes its nutrients are added to the forest soil.

These components of a healthy forest are so important that wildlife trees are required by law in every managed forest in Oregon. The Oregon Forest Practices Act, passed by the state legislature in 1971, was the first in the nation to regulate forest operations on private and state land. The Act protects fish and wildlife, sets standards for reforestation, and lessens impacts of logging, road building, and other operations on soil, water, and air. As scientists learn more about the intricate workings of a forest, the Forest Practices Act continues to be revised.

A FOREST COMMUNITY

By following nature's lead, Oregon's new forests continue to be inviting to people as well as animals. The clear, clean water that fish and beaver enjoy also draws hikers, picnickers, and fishermen. Mushrooms, huckleberries, and other wild edibles are savored by human visitors as well as animal residents. And campers and backpackers sometimes stay in the forest overnight, sleeping nestled amongst the trees like any other creature of the woods.

Many animal species make their homes in second growth forests, including deer, elk, grouse, bear, mice, rats, and rabbits. Different animals depend on different stages of forest growth for food and shelter, habitats which change as the forest matures through its cycle. While some species prefer older forests, others thrive in second growth, drawn to the vigorous new plant life in young forests. Some animals can live in either habitat.

The benefits of Oregon's managed forests extend beyond the well-fed elk and bear. Many communities in Oregon, including some in major metropolitan areas, draw their water supply from second growth forests. And myriad wood and paper products from Oregon's woods are used daily throughout the United States and in other parts of the world.

*Second generation ponderosa pine along the East Fork of
Beech Creek, near John Day, Grant Country.*

Our state is one of the best places in the world to
grow trees. This fact has shaped Oregon's landscape,
history, and the people who live here. And though
it has changed through time, the bond between
Oregon's people and forests continues today. To feel
it, just take yourself for a walk in the woods. And
while you are there, open your senses to the forest
and see, hear, smell, and feel all you can.

— Patricia K. Lichen
Oregon City

*A great horned owl lives and hunts in a Northeast
Oregon pine forest managed by Boise Cascade.*

*A young Douglas-fir grows out of an old growth conifer stump
on Larch Mountain, Multnomah County.*

Wild rhododendrons bloom on a logged hillside,
surrounded by second growth Douglas-fir and seedlings.

Freres Lumber Co., Inc.
Near North Fork Santiam River

Marion County
Photographed June 1990

Ponderosa and lodgepole pine near La Pine, Deschutes County.

Vine maple changing colors in autumn amid a forest of Douglas-fir,
Warm Springs Indian Reservation, Jefferson County.

*Fallen vine maple leaves, mosses, and lichens
decorate a burnt tree root, Lane County.*

Pearly everlasting encircles young hemlock and noble fir in a misty coastal forest.

Stimson Lumber Co.
Area harvested 1940s
Natural regeneration, planting, aerial seeding

near Elsie
Clatsop County
Photographed September, 1995

Assorted hardwoods add fall color to a stand of second growth Douglas-fir.

Lone Rock Timber Co.
Timberland

near Drain
Douglas County
Photographed September, 1994

A logged stump provides a nourishing habitat for
Pacific yew seedlings, Multnomah County.

*An old springboard stump is home to mosses
and sword ferns, near Dee, Hood River County.*

Clockwise from top left:
Douglas-fir branches; fallen ponderosa pine cones
and needles; ponderosa pine needles; western redcedar branches.

A young western hemlock grows in a second growth Douglas-fir forest.

Willamette Industries
Logged 1932 after wildfire; natural regeneration
Commercially thinned by horse logging, 1975
Fertilized 1981, 1991

Crabtree Valley
Linn County
Photographed October, 1993

The Siletz River meanders through second growth forest, near Siletz, Lincoln County.

Second growth Douglas-fir and hemlock rise above sword ferns, near Lacomb, Linn County.

*Pink and yellow monkeyflowers line
Clear Branch Creek, Hood River County.*

A stand of sun-dappled aspen like this can be threatened by varying water tables,
or by grazing cattle, deer, and elk, which feed on the tender shoots.

Near Aspen Lake
Klamath County
Photographed May, 1994

Douglas-fir emerge in a grassy meadow
near Table Rock, Clackamas County.

Unusual growth of Douglas-fir probably caused by
heavy snow pack when the tree was young.

Willamette Industries
Logged in 1932 after a wildfire
Natural regeneration; never thinned
Fertilized 1981, 1991

near Crabtree Valley
Linn County
Photographed October, 1993

*A lodgepole pine seedling rises from the center
of a dead bitterbrush, Deschutes County.*

A stand of ponderosa pine in the high desert, west of Winter Ridge.

Weyerhaeuser Co.
Stand acquired 1910-1937
Periodically selectively harvested
Natural regeneration

Near Summer Lake
Lake County
Photographed Fall, 1992

*Autumn-golden larch stand among a slope of ponderosa pine
on Boise Cascade timberland, Northeast Oregon.*

Fall colors line the Sprague River, Klamath County.

*A young evergreen against a background
of changing vine maple, Multnomah County.*

Moss-covered alder and lush forest greenery, Clackamas County.

Autumn colors reflected in the Umpqua River, Douglas County.

Bracken ferns and a decaying log bathed by evening light, Douglas County.

A quartet of young dark-eyed juncos demand their next meal.

A snag provides shelter for cavity-dwellers on a steep,
fog-shrouded slope of mixed conifers, Hood River County.

*Old Man's Beard, a lichen commonly found growing on tree bark or branches,
festoons dogwood and ash trees near Rogers Mountain, Linn County.*

A smooth-barked madrone in the Siskiyou Mountains.

Rough & Ready Lumber Co.
Timberland

Illinois Valley
Josephine County
Photographed September, 1995

Black bear, deer, and elk make their homes in this hillside of Douglas-fir and red alder.

Georgia-Pacific
Harvested 1954–1957
Planted & seeded 1958, 1959
Commercial thinning began 1995

from the slopes of DiamondPeak,
Lincoln County

*Singing waters tumble among moss-covered boulders
along Still Creek, Mt. Hood National Forest.*

A grassy track leads into a thick stand of alder.

Tillamook State Forest
Burned in series of fires, 1930s–1940s
Although the forest was replanted with
conifers, these alder regrew naturally.

Highway 6 near Jordan Creek
Tillamook County
Photographed February, 1988

Clockwise, from upper left:
A checkerspot butterfly lights on Indian paintbrush;
a ladybug climbs a noble fir trunk; a garter snake slithers
through fallen leaves; a tiny Pacific treefrog
rests on a shaggy mane mushroom.

Flowers bloom in a logged meadow, Clackamas County.

Lupines burst with color on cleared land near Sandy, Clackamas County.

Young spruce trees sprout amid wild foxglove
on the slopes of a logged area on the Coast Range.

Georgia-Pacific   *along Seven Devils Road*
Harvested 1991   *Coos County*
Replanted winter of 1991-1992   *Photographed June, 1995*

Old growth hemlock and Douglas-fir tower above second growth trees.

Willamette Industries
Harvested 1940; natural regeneration
Precommercial thinned 1973
Fertilized 1975, 1981

Crabtree Valley
Linn County
Photographed October, 1993

Saplings share hillside with wildflowers below a "landing,"
where harvested trees were gathered for shipment to a mill.

Georgia-Pacific
Harvested 1991
Replanted winter 1991–1992

along Seven Devils Road
Coos County
Photographed June, 1995

Lodgepole pine seedlings thrive in the fertile ground after a forest fire near Sunriver.

Deschutes National Forest
Photographed May, 1993

*In an open area of mixed forest, vine maple, huckleberries,
and snowberries turn to fall colors, Marion County.*

Lichen, mosses, Douglas-fir needles and cones carpet a forest floor, Clackamas County.

False hellebores emerge in meadow in the Ochoco Mountains.

Second growth forest of Douglas-fir, ponderosa pine, and larch provides a scenic backdrop as the Middle Fork John Day River flows past ranch land in the Blue Mountains, Grant County.

Snow flocks the branches of a young Douglas-fir.

Boise Cascade   *near Bailey Butte*
Planted 1986   *Jackson County*
                                              Photographed January, 1991

Looking down on snow-covered conifers lining Mill Creek,
Warm Springs Indian Reservation, Wasco County.

Sunlight highlights a big leaf maple along the South Fork of the McKenzie River, Lane County.

Penstemon and paintbrush near Elkhorn, Marion County.

Ponderosa pine rise behind white-rayed mule's ears on the edge of Big Summit Prairie.

Selectively harvested 1960s, 1970s
Natural regeneration

Ochoco Mountains
Crook County
Photographed May, 1988

*Replanted Douglas-fir compete with rhododendrons for sunlight
in harvested area near Marcola, Lane County.*

A weather-worn tree root surrounded by fall-colored vine maple, Clackamas County.

*Big leaf maple along Crabtree Creek, a riparian area surrounded by a forest stand
that was harvested and replanted in 1984. The stream is home to cutthroat
and rainbow trout, and a run of steelhead.*

Willamette Industries
Harvested 1984
Replanted 1984

Crabtree Valley
Linn County
Photographed October, 1993

Clockwise, from top left:
Evergreen violets emerge between fallen
spruce cones; bunchberry changes to fall color;
Oregon-grape amid patches of late spring snow;
a larch branch catches fallen leaves
of vine maple and cottonwood.

Turning vine maple among mature second growth Douglas-fir.

Longview Fibre Co.
Mid-Columbia Tree Farm
Acquired early 1980s
Thinned 1993; fertilized 1994

Near West Fork of Hood River
Hood River County
Photographed October, 1994

Old Man's Beard drapes pond-side conifer branches, Deschutes County.

A nurse log gives new life to young trees in Nehalem Bay, Tillamook County.

*Mosses flourish in the exposed root system of a downed log
in a harvested area near Lowell, Lane County.*

*Huckleberries and vine maple spring from an old growth stump
on Larch Mountain, Multnomah County.*

*Rhododendrons bloom in a young plantation of Douglas-fir at
Lolo Pass, Mt. Hood National Forest, Clackamas County.*

*In a harvested area adjacent to a stand of old growth, young Douglas-fir
mingle with vivid vine maple, Mt. Hood National Forest.*

*A free-standing snag stands sentinel before a hillside
of young Douglas-fir, Clackamas County.*

Paintbrush and wildflowers carpet the approach to a stand of mixed-age ponderosa pine.

Selectively harvested 1960s, 1970s
Natural regeneration

Ochoco Mountains
Crook County
Photographed May, 1990

*Douglas-fir and hemlock line a road into the fog-shrouded
forest near Green Mountain, Linn County.*

The setting sun silhouettes a Western Oregon forest scene, Lane County.

*Bigleaf maple leaf caught in the bark
of a Sitka spruce, Clatsop County.*

About the Photographer

Portland native Steve Terrill is one of Oregon's finest landscape
photographers. His images have been published in numerous books,
magazines, and calendars, including *National Geographic, National
Wildlife, Outdoor Photographer, Travel & Leisure, Audubon,* and Sierra Club
publications. His first book, *Oregon: Images of the Landscape* was published by Westcliffe in 1987, followed by *Oregon: Magnificent Wilderness*
(1991), *Oregon Coast* and *Oregon Wildflowers* (1995), and three annual
Oregon calendars. To collect the images in *Oregon's New Forests*, Steve
visited private and public forest lands throughout the state.

About the Oregon Forest Resources Institute

OFRI was established by the Oregon Legislature in 1991 to "enhance
and support Oregon's forest products industry" by improving public
understanding of forest practices and products, and by encouraging
sound forest management.